The Marines' Hymn

From the Halls of Montezuma,
To the shores of Tripoli;
We fight our country's battles
In the air, on land, and sea;
First to fight for right and freedom
And to keep our honor clean;
We are proud to claim the title
Of United States Marine.

Our flag's unfurled to every breeze
From dawn to setting sun;
We have fought in every clime and place
Where we could take a gun;
In the snow of far-off Northern lands
And in sunny tropic scenes;
You will find us always on the job
The United States Marines.

Here's health to you and to our Corps
Which we are proud to serve;
In many a strife we've fought for life
And never lost our nerve;
If the Army and the Navy
Ever look on Heaven's scenes;
They will find the streets are guarded
By United States Marines.

For Crystal, Austin and Graham. - Brandon Barnett

For my wonderful family. - Rebecca Wochner

A Salute To Our Heroes - The U.S. Marines

Copyright © 2010 by B.W. Barnett

**Officially Licensed By
The U.S. Marine Corps**

Requests for permission to make copies of any part of the work should be submitted online at info@mascotbooks.com or mailed to Mascot Books 560 Herndon Parkway #120, Herndon, VA 20170

Printed in the United States.

PRT0410A

ISBN-13: 978-1-936319-00-8
ISBN-10: 1-936319-00-4

Mascot Books
560 Herndon Parkway #120, Herndon, VA 20170

www.mascotbooks.com

Have a book idea? Contact us at:
Mascot Books
P.O. Box 220157
Chantilly, VA 20153-0157
info@mascotbooks.com

A Salute To Our Heroes
THE U.S. MARINES

Brandon W. Barnett
Illustrated by Rebecca Wochner

Meet Chesty the bulldog.
Chesty is the mascot of the U.S. Marine Corps.
Read along as Chesty teaches us about the
Marines and the important job they do.

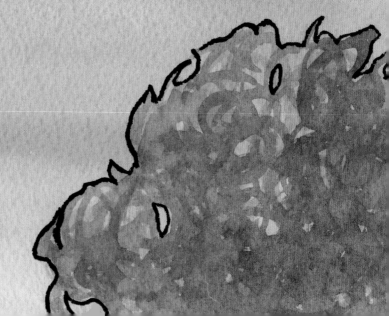

From all over the country,
Men and women decide,
To join the Marine Corps
– To be tested and tried.

Marines first go to boot camp,
Where drill instructors are loud.
If they pass, they become
One of "the few and the proud."

Training is important to make Marines better.
Marines train all the time, even in bad weather.
Marines do pull-ups and crunches and go on long runs...

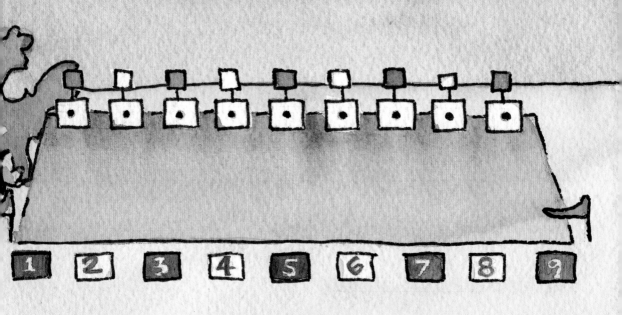

And they must always shoot straight
When they fire their guns.

Marines take an oath,
To "support and defend."
To the country we love,
There is no better friend.

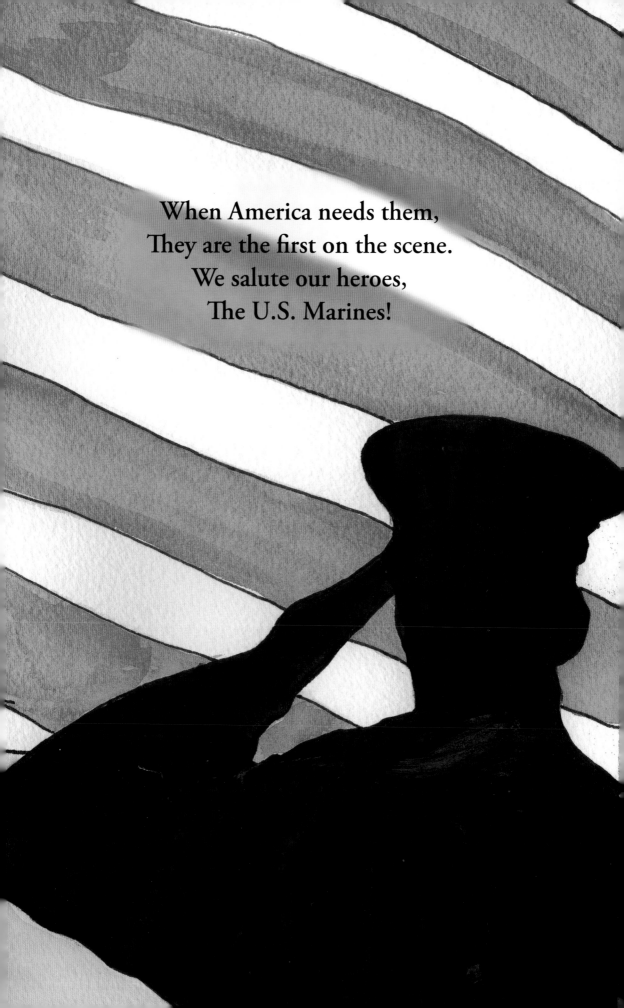

When America needs them,
They are the first on the scene.
We salute our heroes,
The U.S. Marines!

Sometimes Marines go to far away lands,
And live in the jungle or in the hot sand.

They leave their families and friends behind,
To go forward and serve wherever assigned.

Marines wear camouflage
Made of tan or green.

It helps them hide,
So they cannot be seen.

The Navy helps out
When Marines need to take trips.
They give Marines a ride
On their really big ships.

Marines ride on these ships
From sea to sea.
When on land
Marines use tanks or humvees.

Some Marines fly helicopters
High in the sky,
And others fly jets
That *zoom* right by.

Radio men on the ground
Are doing their share,
To talk to the pilots
Up in the air.

After gathering their gear
And lacing their boots,
Marines march many miles
With the rest of the troops.

But when marching cannot get them
From here to there,
Marines grab a parachute
And arrive from the air.

Marines are great friends,
Some even say brothers.
They watch out for danger
And protect one another.

In the back of big trucks
Marines ride down the trail.
They are Semper Fidelis.
Marines never fail.

When Marines return home
From serving abroad,
People gather to welcome them
– To cheer and applaud.

They have done a great job
Protecting you and me.
We salute our heroes,
The U.S. Marines!

Marine One flies the President
When he has somewhere to go.

At ceremonies,
The Silent Drill Platoon steals the show.

Whether earthquake or storm,
Conflict or war,
Marines are up to the task
And ready for more.

Marines work together
As one big team.
We salute our heroes,
The U.S. Marines!

<u>Glossary</u>

Boot Camp - the 12-week school where young men and women learn how to be Marines.

Camouflage - the use of color and objects that help Marines blend into their surroundings.

Drill Instructor - the teachers at boot camp that help new Marines learn everything they need to know. Drill instructors are known for yelling and screaming so that everyone can hear them.

Humvee - a vehicle that looks similar to a Jeep. A Humvee typically has a machine gun mounted to the top and is used by Marines for many operations.

Marine One - the name of the Marine Corps helicopter that flies the President of the United States.

Oath - a very important promise. Marines take an oath to protect America.

Oorah - the spirited battle cry of the Marines.

Salute - to pay honor and respect by raising your right hand to the tip of your hat (cover).

Semper Fidelis - latin phrase meaning "always faithful;" the motto of the U.S. Marines.

The Few and The Proud - a title given to Marines because they stand out as the most disciplined and dedicated of the armed forces.

Training - for Marines, training includes working out to stay physically fit, reading and studying books to sharpen their minds, and practicing military skills, like shooting.

U.S. Marine Corps - the elite branch of America's armed forces. There are more than 202,000 active Marines spread across the United States, Japan, and many other parts of the world.

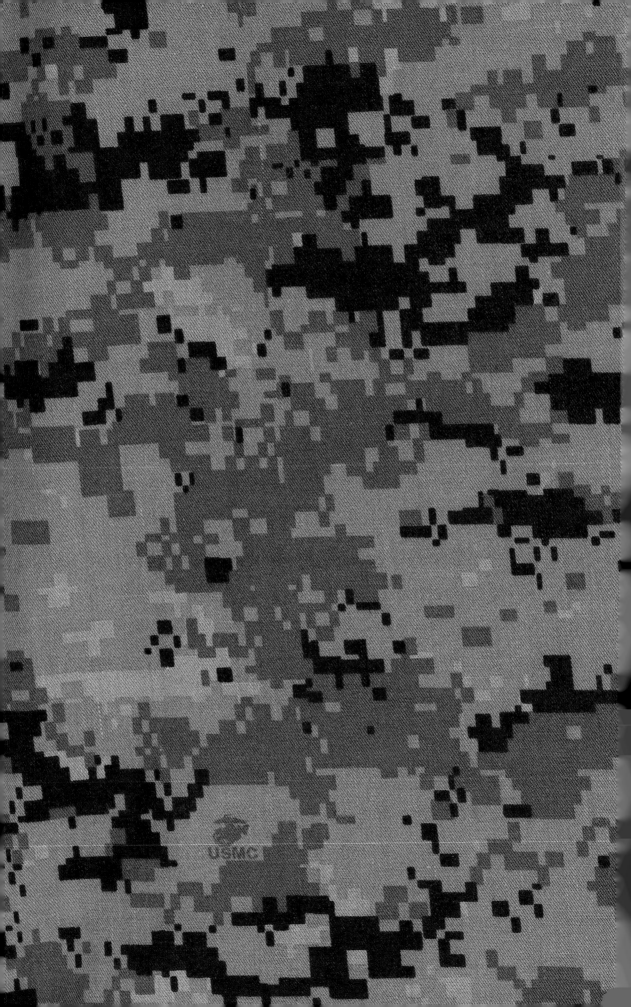